SAE EDGE™
RESEARCH REPORT

Unsettled Topics on the Use of IVHM in the Active Control Loop

Rhonda D. Walthall
Collins Aerospace

EDGE DEVELOPMENT TEAM

Dr. Shreeder Adibhatla, *General Electric Aviation*
Carina Birgersdotter, *SAAB, retired*
Andy Hess, *The Hess PHM Group*
Dr. Ian Jennions, *Cranfield University*
Dave Larsen, *Collins Aerospace*
Linda O'Flarity, *Pratt & Whitney*
Steve O'Flarity, *Pratt & Whitney*
Dr. Ravi Rajamani, *drR2 consulting*
Douglas Felipe Rodrigues da Silva, *Embraer S.A.*
Dr. Antonios Tsourdos, *Cranfield University*
Dr. George Vachtsevanos, *Georgia Institute of Technology, retired*

SAE INTERNATIONAL®
Warrendale, Pennsylvania, USA

About the Publisher

SAE International® is a global association of more than 128,000 engineers and related technical experts in the aerospace, automotive and commercial-vehicle industries. Our core competencies are life-long learning and voluntary consensus standards development. Visit sae.org

SAE EDGE™ Research Report Disclaimer

SAE EDGE™ Research Reports focus on topics that are dynamic, in which knowledge is incomplete, and which have yet to be standardized. They represent the collective wisdom of a group of experts and serve as a practical guide to the reader in understanding unsettled subject matter. They are not meant to provide a recommended practice or protocol. The experts have assembled as a community of practitioners to contribute and collectivize their thoughts and points of view; these are not the positions of the institutions or businesses with which they are affiliated, nor is one contributor's perspective advanced over other contributors. SAE EDGE™ Research Reports are the property of SAE International and SAE alone is responsible for their content.

About This Publication

SAE EDGE™ Research Reports provide state-of-the-art and state-of-industry examinations of the most significant topics in mobility engineering. SAE EDGE™ contributors are experts from research, academia, and industry who have come together to explore and define the most critical advancements, challenges, and future direction in areas such as vehicle automation, unmanned aircraft, cybersecurity, advanced propulsion, advanced manufacturing, Internet of Things, and connectivity.

Related Resources

SAE MOBILUS® Automated & Connected Knowledge Hub
https://saemobilus.sae.org/automated-connected/

SAE Team

Frank Menchaca, Chief Growth Officer
Michael Thompson, Director, Standards, Information and Research Publications
Monica Nogueira, Acquisitions Director
Beth Ellen Dibeler, Product Manager
William Kucinski, Managing Technical Editor

Printed in USA

EPR2020011
ISSN 2640-3536
e-ISSN 2640-3544
ISBN 978-1-4686-0185-5

To purchase bulk quantities, please contact: SAE Customer Service

E-mail: CustomerService@sae.org
Phone: 877-606-7323 (*inside USA and Canada*)
+1-724-776-4970 (*outside USA*)
Fax: +1-724-776-0790

https://www.sae.org/publications/edge-research-reports

About the Editor

Rhonda D. Walthall, photo by SAE International

Rhonda D. Walthall is a Technical Fellow at Collins Aerospace in Charlotte, NC, a division of Raytheon Technologies Corporation (RTX). In her role, she focuses on Design for Prognostics and Health Management (PHM) and Digital Thread initiatives. She is an industry-recognized leader in the development of standards and best practices for PHM and Integrated Vehicle Health Management (IVHM).

Walthall earned her Bachelor of Science Degree in Aeronautical and Astronautical Engineering from Purdue University and her Master's degree in Business Administration from Pepperdine University. She began her career as a Flight Test Engineer for the McDonnell Douglas Aircraft Company before working for Northwest Airlines as a Powerplant Engineer. In 2003, she joined Hamilton Sundstrand and has held positions of increasing responsibility, leading to her role today as a Technical Fellow.

Walthall is a member of the SAE International Board of Directors and is the Vice President of the PHM Society Board Directors. She has been actively engaged in SAE standards development and leadership since 2004. She was recognized by SAE as a Top Contributor of the Year (2019) and received the Rodica Baranescu Award for Technical Excellence & Leadership (2018) and the James M. Crawford Technical Standards Award for Outstanding Achievement (2016). She holds three patents and has contributed significantly to numerous SAE documents and authored chapters in SAE and IEEE publications. She has presented technical papers and has participated as a keynote speaker, invited speaker, or panelist at numerous PHM-related conferences.

In 2020, Walthall received the Outstanding Aerospace Engineer Award from Purdue University, where she is a member of the Industrial Advisory Council to the School of Aeronautics and Astronautics.

contents

Atosan/Shutterstock.com

Unsettled Topics on the Use of IVHM in the Active Control Loop

Abstract

The growth in global economies has led to a world that has become much more mobile in the last few decades. The number of enplanements has increased and is expected to continue to do so at an annual average rate of 1.8% through 2039 [1]. Prior to the COVID-19 pandemic, the number of aircraft in service was expected to increase annually to meet the travel demand. Next-generation, more-complex aircraft were scheduled to replace the older aircraft at a pace that still allowed sufficient capacity to meet the increasing demand. The events of 2020 have driven the industry to accelerate retirement of older aircraft while deferring the introduction of new aircraft. While the length of the industry recovery period cannot be predicted, most analysts believe that demand for travel will return once a vaccine is widely available. The impact to the design of next-generation aircraft will likely be shaped by technologies that are being accelerated for the post-COVID world as well as for new mobility platforms. Technologies, such as artificial intelligence and fault-tolerant and self-adapting control, will use integrated vehicle health management (IVHM) capabilities as part of the decision-making processes. This SAE EDGE™ Research Report seeks to explore the unsettled issues surrounding embedding IVHM information into the active control loops of modern aircraft systems and in future generations of aircraft designs.

NOTE: SAE EDGE™ Research Reports are intended to identify and illuminate key issues in emerging, but still unsettled, technologies of interest to the mobility industry. The goal of SAE EDGE™ Research Reports is to stimulate discussion and work in the hope of promoting and speeding resolution of identified issues. SAE EDGE™ Research Reports are not intended to resolve the challenges they identify or close any topic to further scrutiny.

RHONDA D. WALTHALL
Collins Aerospace

Edge Development Team

Dr. Shreeder Adibhatla, *General Electric Aviation*
Carina Birgersdotter, *SAAB, retired*
Andy Hess, *The Hess PHM Group*
Dr. Ian Jennions, *Cranfield University*
Dave Larsen, *Collins Aerospace*
Linda O'Flarity, *Pratt & Whitney*
Steve O'Flarity, *Pratt & Whitney*
Dr. Ravi Rajamani, *drR2 consulting*
Douglas Felipe Rodrigues da Silva, *Embraer S.A.*
Dr. Antonios Tsourdos, *Cranfield University*
Dr. George Vachtsevanos, *Georgia Institute of Technology, retired*

ISSN 2640-3536

Introduction

State of the Industry

Integrated vehicle health management (IVHM) is defined in SAE ARP6803™ as "the unified capability of a system of systems to assess the current or future state of a member system's health and integrate that picture of health within a framework of available resources and operational demand" [2]. The IVHM capability comprises five functional blocks: sense, acquire, transfer, analyze, and act (SATAA). The first three functional blocks leverage data sensing, acquisition, and processing aptitudes that are inherent in the design of complex systems. The analyze and act functional blocks are the additional value that the IVHM capability provides. Specifically, IVHM algorithms provide insight to the current performance (state detection), the current health assessment (normal, degraded, faulty), and the future health state or performance (failure mode), and also an advisory alert for when action should be taken to avoid a functional failure of the component or system, within a timeframe of uncertainty. Figure 1 illustrates the functional blocks of an IVHM system as it relates to the open system architecture for condition-based maintenance (CBM) in ISO 13374.

For IVHM to be effective, algorithms must "accurately predict the time to failure or Remaining Useful Life (RUL) of critical components or systems, and effectively isolate the root cause of failures once effects are observed, and provide system stakeholders actionable information in support of operations, maintenance and support"[3]. Figure 2 illustrates the timeline in which the value of IVHM can be realized. Functional failure refers to the inability of a component, system, or vehicle to fulfill its design purpose. Functional failure is considered to be the most severe degree of malfunction or degradation, whereas a fault, or an incipient fault, is considered to be the symptom indicating a condition indicative of an impending failure [4]. Maintenance, repair, and overhaul (MRO) refers to the opportunity to perform the corrective action (i.e., servicing, repairing, or removing and replacing the failing component or system) to mitigate the risk of functional failure.

In the aviation industry, IVHM capability is currently comprised of onboard and off-board elements. The onboard elements include hardware (e.g., physical sensors, data recorders, aggregators, transmission devices) and software (e.g., built-in test (BIT), logic for capturing, processing, manipulating, and transmitting data, virtual sensors, model-based diagnostics and prognostics). Many of the onboard elements exist primarily for the safe operation and control of the aircraft and to support time-sensitive information needed for fault detection, isolation, and accommodation (FDIA). The off-board elements include the data ingestion process (receiving, processing, and storing of data), the data analysis process (trending, feature extraction, diagnostic/prognostic algorithms and reasoners), and the alert generation process (recommendations for maintenance or operational action). Traditionally, the IVHM capability has been used to support maintenance, which occurs after the aircraft lands. Many times, the maintenance performed is based upon what is detected and accommodated during the flight; other times, the maintenance performed is based on what is predicted by the IVHM system.

Tomorrow's aircraft will have increasing levels of autonomy, which means that the onboard control systems

FIGURE 1. Open system architecture for condition-based maintenance layered reference model.

OSA-CBM Functional Blocks	SAE/IVHM PHM Stages
Data Acquisition (DA)	Sense
Data Manipulation (DM)	Transfer
	Acquire
State Detection (SD)	Analyze
Health Assessment (HA)	
Prognostics Assessment (PA)	
Advisory Generation (AG)	Act

FIGURE 2. IVHM value timelines.

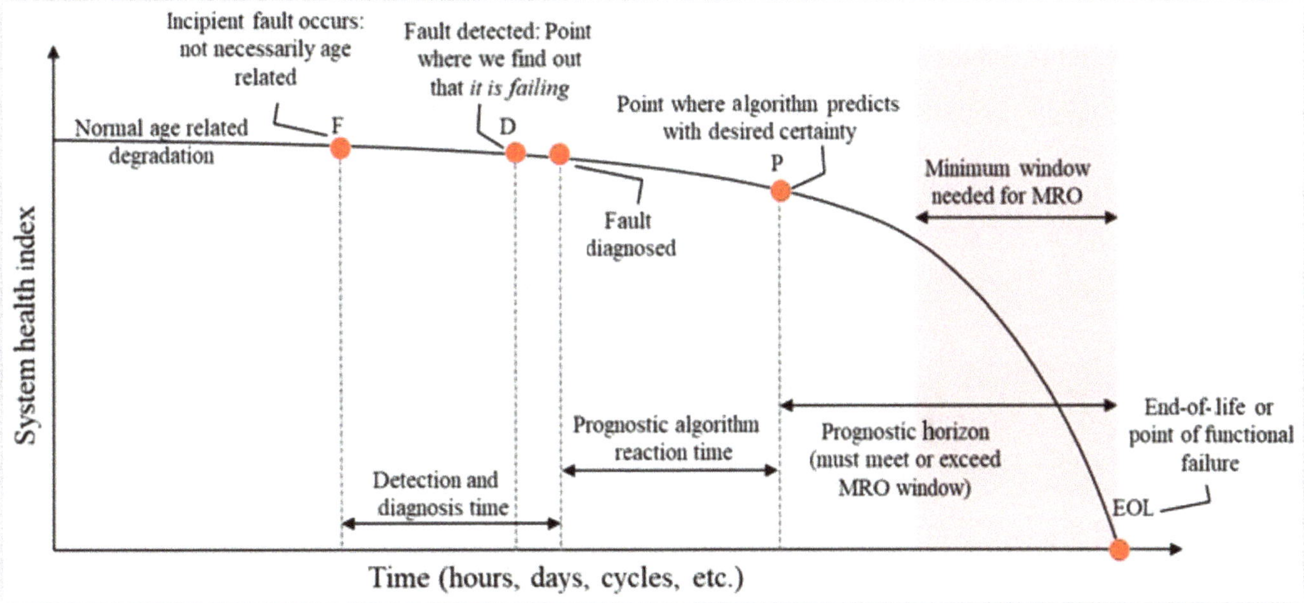

must be able to use information differently to ensure the autonomous systems work together safely. The IVHM capability levels support the levels of increasing autonomy and are shown below [5]:

- Level 0: Limited to on-aircraft warning indications
- Level 1: Enhanced onboard diagnostics
- Level 2: Real-time data available for off-board diagnostics and prognostics
- Level 3: Onboard prognostics at the component level
- Level 4: Onboard prognostics at the system level
- Level 5: Self-adaptive, diagnostics and prognostics integrated into aircraft controls

Today's aircraft have IVHM capabilities that are consistent with Level 2. Some military aircraft have achieved Level 3 to a limited extent. Fully autonomous aircraft will need to achieve Level 5.

Active control is the process of using data from physical or virtual sensors for regulation and feedback of continuously operated, dynamic systems on the aircraft. Because of the potential impact to safety, critical systems have two or more fully redundant and dissimilar control systems that run simultaneously with redundant sensors and redundant software to provide real-time information in case one data path fails. Fault-tolerant control systems are capable of accommodating component and system failures automatically. Therefore, FDIA is embedded into the active control of most safety-critical systems, such as propulsion systems.

The integration of IVHM capability into active control systems of civil transport aircraft has been confined to the state detection and the health assessment (i.e., fault detection) for good reason: system integration constraints have prevented information from being shared between systems, the IVHM elements have not been designed to meet the certification requirements for airborne equipment, the nondeterministic nature of the prognostic algorithms has been undesirable, and the historical mandate for a pilot to have full authority over aircraft systems has limited the expansion of autonomy.

The integration of IVHM capability into the control system of rotorcraft has not been as confined as for civil transport aircraft. The embodiment of an IVHM system on rotorcraft is a health and usage monitoring system (HUMS). The term HUMS is believed to have originated in the oil and gas industry in 1986, following the crash of a Boeing 234LR Chinook helicopter in the North Sea. The high and intolerable accident rate in the North Sea [6] drove the industry to develop consensus standards around HUMS and to adopt varying degrees of HUMS capability. In 1999, the United Kingdom's Ministry of Defence issued a mandate to adopt HUMS. The Federal Aviation Administration (FAA) and European Union Aviation Safety Agency (EASA) issued guidance for airworthiness approval for installing a HUMS capability and instructions for continued airworthiness using the HUMS capability and provided guidance for the use of HUMS to obtain maintenance credits [7]. HUMS technologies advanced through increased data integration and interoperability, improvements in sensors and data storage, and the proliferation of diagnostic and prognostic techniques. The high volume of research into vibration monitoring and feature detection provided proven methods that increased safety, further leading to the widespread adoption of HUMS. In 2008, the US Army issued a mandate to transition away from traditional CBM to a proactive fleet maintenance strategy referred to as CBM+, which required the installation of HUMS on their aircraft.

On some modern rotorcraft today, the information provided by the HUMS is used by the flight management computer (mission computer on military aircraft) to modify flight control parameters to ensure the flight can be completed safely. If the rotorcraft is used for nonaggressive operations, such as passenger transportation or medical evacuation, the exposure to high-stress loads is much lower than during aggressive military operations. Hence, there is more leeway in how the rotorcraft can be operated and potentially restricted. Specifically, the power output to the rotor can be constrained if the HUMS detects a degraded condition. By limiting the power output to the rotor, the HUMS is able to lower rotor and engine speed and limit the G-forces that can be commanded by the pilot. The result is a more conservative operation, well within the certified flight envelope. Depending on the severity of the degradation, the pilot may not be aware of the constraints imposed by the flight control system based on inputs from the HUMS.

Integration of the IVHM capability into active control systems on military aircraft is consistent with civil transport aircraft and rotorcraft for similar aircraft types. For military aircraft that are dissimilar, such as the Lockheed Martin F-35 Lightning II and unmanned aerial vehicles (UAVs), the implementation strategy has been very different. For example, the strategy for the F-35 was to perform all of the health-monitoring functionality onboard the aircraft in either the engine controller or the mission computer. While this strategy was not fully realized, the vision led to extensive research into information fusion and the advancement of diagnostic and prognostic technologies by universities, government agencies, and industry partners. For the purposes of this SAE EDGE™ Research Report, military UAVs are not discussed.

For aircraft such as the F-35, fault-tolerant adaptive control and system integration are key to safe operations; in essence, the active control systems must keep the pilot safe and the mission going. The active control systems must be capable of detecting and accommodating imminent faults and predicting if the fault progression could result in a catastrophic situation. Despite the focus on FDIA for safety, onboard software on military aircraft is similar to civilian aircraft in that not a lot of focus has been placed on onboard aircraft and engine fault correlation. While some correlations are known, many are not. During the design and development of engine control systems, emphasis is placed on understanding how the engine will respond to a fault, not on how the aircraft will respond. As a result, much of fault correlation still occurs post-flight on the ground. In this particular scenario, an onboard IVHM capability could correlate the faults from all systems, and then either provide the information to the pilot or inform the control system to take immediate action, depending on the severity of fault consequences.

System integration is complex and challenging. Integrating the engine control software within the aircraft control system software is very limited today. Integrating pilot responses to annunciated fault conditions into the aircraft control system, such that appropriate action is taken, will require significantly more integration between systems and subsystems, and will require experts from different companies that understand the entire air vehicle system to work together closely.

Should IVHM capabilities be embedded in the active control loops of transport aircraft? What challenges would the industry need to overcome in order to embed IVHM capabilities into the active control loops? A community of experts was interviewed, and this report reflects their thoughts regarding the pros and cons, the challenges, and the way forward on these unsettled topics.

Challenges to Implementing IVHM Capabilities into Active Control Systems

Historically, IVHM has been used for economic purposes on civil transport aircraft: to lower the cost of ownership by reducing the impact of maintenance-related disruptions, and by optimizing the maintenance performed. Since IVHM has been considered an economic influencer rather than a safety influencer, many in the civil transport industry do not believe the benefits derived from implementing an onboard IVHM capability would outweigh the cost of installing the hardware and software to enable the capability. (This position differs in the rotorcraft industry.) Additionally, the industry has resisted implementing new control designs due to the high cost of certification. Optimization studies of control systems have not been widely practiced when identifying the requirements for a new aircraft. As a result, aircraft manufacturers have not always implemented good features from previous aircraft designs but have instead focused on meeting the burden of certification by keeping things common. Having said that, the benefits of IVHM are slowly being recognized in the commercial world and health management features are being incorporated into the aircraft, such as extensive vibration monitoring, oil debris monitoring, use of engine health management (EHM) systems to assess extended operations (ETOPS) viability, and so on.

Today's transport aircraft are very diverse in terms of the level of IVHM implementation. The implementation has increased over the years as newer generation aircraft have entered service with advanced technologies that enabled more data collection and storage and more embedded algorithms for fault detection, diagnosis, and prognosis. In general, the degree of IVHM implementation on a transport aircraft falls under one of four levels, as described below.

The first level of implementation can be thought of as a "dark" aircraft. For older regional aircraft and large (single-aisle) transport aircraft, such as the Boeing 737 series and the A320 family, the onboard sensor suite and the limited central data acquisition system have greatly hindered the ability to implement an IVHM capability. Some of these older aircraft lack a central maintenance function to consolidate fault messages onboard. Even though most systems are capable of detecting faults through BIT, the fault isolation and diagnostics are handled post-flight, on the ground by a human. The engine manufacturers were able to develop EHM capabilities by using fully redundant sensors and partitioned controllers and by transmitting data from the controllers to an onboard data acquisition device for manual or automated download from the aircraft. As such, EHM capabilities continuously advanced to include onboard FDIA and off-board prognostics. Integrating IVHM information into the active control of any system other than the propulsion system would not be feasible on these aircraft with the existing data acquisition and transmission technologies available. Newer data acquisition and transmission technologies hold promise for aftermarket retrofit of, or at least partial or targeted IVHM on these aircraft, if they can be shown to be economically viable.

The second level of implementation came about when the central maintenance computer (CMC) functionality was first introduced on the Boeing 747-400 [8]. This functionality is now basic equipment on all transport aircraft. The CMC functionality consolidates system status and detected faults and sends that information to an onboard communications device, which transmits the data to a ground-based analysis system. Once on ground, a human examines the faults and sensor readings, and makes a recommendation for maintenance. Even though the fault consolidation occurs onboard, the aircraft systems are not integrated. Hence, a fault in an upstream system may impact a downstream system; however, the upstream fault may not be detectable until the off-board fault analysis is performed. While the CMC functionality aids in diagnostics, the functionality does very little to advance prognostics. Some aircraft are equipped with separate EHM units installed on the engines to capture information needed for FDIA, engine condition monitoring, and off-board prognostics. Retrofit solutions that tap into the existing electronic data busses to record and transmit data are available but are constrained by the limited sensor suite available. Therefore, integrating IVHM capabilities into the active control of systems other than the propulsion system would be feasible but highly limited.

The third level of implementation coincided with mandates from the regulatory agencies to implement flight data monitoring programs. The operators exchanged traditional aircraft with new, more efficient, ETOPS-approved aircraft that had high-thrust engines with advanced, fully redundant control systems. When the Boeing 777 entered service in 1995, ETOPS approval was made possible in part due to the Airplane Information Management System, which was the most advanced and integrated data management system ever undertaken on a new aircraft design [9]. The data management system and the expanded sensor suite enabled health-monitoring capabilities to be developed for other components and systems, such as the auxiliary power unit (APU) and the environmental control system (ECS). Off-board diagnostic and prognostic algorithms were generated using snapshot reports and full-flight data recordings. Despite the onboard data management system, the onboard systems remained highly independent, forcing much of the data integration to take place in the off-board IVHM system. On ground, humans analyze the alerts generated by the IVHM systems to make informed decisions for maintenance. Retrofit solutions that increase the level of onboard data aggregation are available for aircraft via Supplemental Type Certificate (STC) approval and are economically viable for these aircraft. The availability and integration of data would support the implementation of IVHM information into the active control of targeted components and systems if the IVHM elements were demonstrated to meet the hardware and software certification requirements.

The fourth level of implementation refers to the latest generation of modern aircraft, capable of generating big data, such as the Airbus A350 and A220, Boeing 787, Gulfstream G650, Embraer E2, and advanced military aircraft. The large volume, velocity, and variety of structured and unstructured data available from snapshot reports and full-flight data offered the ability to develop IVHM capabilities for many components and systems. However, the magnitude of the data, as well as data access and ownership issues, have presented new challenges for using the data for IVHM. Since IVHM has been used primarily for maintenance decisions, the method for transmitting the data has been driven by economics. The benefit of streaming large amounts of real-time aircraft data has not been perceived to be worth the cost, forcing the diagnostic and prognostic functions to be performed post-flight, on the ground. The ability to have flexibility to modify algorithms as experience is gained has also driven the IVHM capability to remain off-board: it is much easier and cheaper to deploy these changes on ground than it is to re-certify onboard critical system software. Advanced models and machine learning techniques are used by the IVHM system to analyze the data, to extract features from the processed data, to make predictions and estimations about impending failures and their impact, and to generate advisory messages for operations and maintenance. Because of the high level of data capture and integration, embedding IVHM information into the active control of targeted components and systems would be possible if the uncertainty in the diagnostics and prognostics assessments were managed and demonstrated to meet the certification requirements for software.

If we could integrate IVHM information into the active control systems of current generation transport aircraft, would we want to? The answer is "probably not." The cost of implementation would be too high relative to the benefit that would be realized. When considering future generations of efficient and quiet large transport aircraft, eVTOL aircraft, and autonomous vehicles, the answer is "It depends on the mission."

Large Transport Aircraft

For aircraft intended to transport large numbers of passengers over great distances, the next generation of aircraft and the supporting infrastructure cannot remain stagnant while facing the demands of an increasingly mobile world. According to the FAA's National Forecast, published prior to the COVID-19 pandemic of 2020, the number of passengers traveling worldwide is expected to increase from 244 million in 2018 to 491 million in 2039 (which averages an increase of 3.4% per year) [1]. During this same period, the number of air transport pilots is expected to increase by 25,755 (an average increase of 0.7% per year). This disparity in growth rate is projected to result in a pilot shortage if the industry travel demand returns to pre-pandemic levels. One approach to addressing the anticipated pilot shortage is to design transport aircraft for single-pilot operations with more autonomous systems. Autopilot and auto-throttle are well-known applications of autonomous systems with adaptive control laws that can be overridden by input from the pilot. According to a 2016 report in The *Atlantic*, on a typical domestic flight, 95% of the flying is handled by one of these autonomous systems [10]. Autonomous systems must be able to monitor their own performance, and either automatically reconfigure control laws to achieve a conservative (safer) state or inform the pilot to take appropriate action to mitigate a threat. This means that successful autonomous systems will have some level of diagnostic and prognostic capability embedded in the active control system in order to assess the current condition, predict the future condition, inform the pilot, or reconfigure the active control laws. These autonomous systems will likely have two levels of fault-tolerant reconfiguration: low-level control adaptation at the component level and mid-level control redistribution at the system level. That is, if a failure is detected in a component, the control system governing that component would reconfigure the operating parameters to use alternate laws or sources of information, such as redundant or virtual sensors, to mitigate the risk of component functional failure at the aircraft level. If the low-level control reconfiguration is not sufficient to mitigate the risk of component failure at the aircraft level, then the next higher governing control system would ignore the failed component and redistribute the demand or load to ensure continued, safe operation. An example of this exists today on the Boeing 787, where the electric power system is capable of automatic load shedding and reconfiguration to ensure that critical aircraft systems remain powered at all times.

The following examples suggest opportunities where using IVHM information embedded in the active control could support more autonomy:

- *Enhanced BIT*: Any component or subsystem that performs a self-BIT assessment could have software enhancements that report more than just PASS/FAIL to the active control. The BIT software could be designed to report margins, minimums, maximums, and other relevant information, such as potential to failure, to the active control system at the time the BIT is performed. This information could allow the active control laws to reconfigure or accommodate as needed, based on the enhanced BIT assessment.
- *Flight control system*: An IVHM system could fuse data from the flight planning system, air traffic control, weather, fuel management system, propulsion system, and other onboard systems to reconfigure the control surfaces to ensure stable flight and optimization for flight efficiency and fuel burn. This information could also be used for route optimization as well as for understanding the impact on fuel burn due to air traffic management and weather constraints.
- *Air management system*: An IVHM system could optimize the cabin temperature and humidity for operators who fly in extreme climate conditions. Suppliers have a difficult task in designing an air management system that performs well in high heat and humidity environments as well as in extreme cold environments. The system as a whole is traditionally designed to operate in nominal conditions for the majority of the aircraft's life cycle, with a smaller percentage of life spent in extreme weather conditions. The IVHM information could assist in optimizing the air management system control laws for the specific aircraft's operating environment. For example, if an aircraft primarily operates in a hot and humid climate, the control laws could be altered to enable the air management system to achieve the desired cabin temperature and humidity while on ground or at lower altitudes while maintaining the common control laws in-flight.
- *Masked faults*: An IVHM system could detect impending failures in aircraft systems that are masked when the system is controlled to a commanded state. For example, air management systems are comprised of many mechanical components. As components degrade, the pack control software adapts to ensure the commanded cabin temperature and pressure are achieved and maintained. As a result, the deteriorated state of the components is not known until a failure of one or more of the components occurs, which often happens when the aircraft is fully loaded with passengers or when the aircraft is parked in an extreme weather environment. Integrating information from the IVHM system into the pack control software could enable the controller to be aware of the degrading state and to set a fault notification before the condition results in a system functional failure.
- *Lubrication system*: An IVHM system could improve the process of monitoring the oil condition and quantity for engines and APUs. Much of the monitoring still takes place on the ground through

oil sampling methodologies, such as spectrometric oil analysis, magnetic chip detectors, and oil level sight gauges. Electronic chip detectors, inline debris monitoring devices, and sensors that indicate a pressure change across a filter are installed for fault detection. Once a fault is detected, the aircraft is grounded until the condition is resolved. Embedding the sensor information into the engine or APU active control could enable an early warning of oil quantity loss or oil contamination and could potentially indict the component that is causing the oil contamination.

- *Hydraulic systems*: An IVHM system could monitor the quality of the hydraulic fluid, which is essential to the efficiency and performance of the hydraulic system and electrohydraulic servo-actuators. The IVHM system could enable early warning of the degraded performance of the hydraulic system, allowing for manual or automated reconfiguration of the flight control surfaces to ensure safe and stable operation.
- *Landing systems*: An IVHM system could provide information to a distributed control system to ensure proper braking is applied based on brake wear, brake temperature, tire pressure, and strut health. Landing systems operate in harsh environments and are typically unable to connect to other aircraft systems. Hence, today's aircraft are hindered by the lack of onboard, real-time integration of landing system sensor data with onboard data acquisition systems. Diagnostic and prognostic assessments are made post-flight once the data has been manually downloaded from the aircraft. A distributed control system that uses a network to allow the landing gear sensors, actuators, and brake controllers to communicate with each other could also include vehicle-level diagnostic and prognostic reasoners (software that is able to infer logical consequences from a set of asserted facts or axioms[1]) to ensure proper braking pressure is applied during all-weather conditions, airport and runway conditions, and pilot-induced force. This situation is mitigated to an extent in some modern aircraft today that have electric brake actuation systems, where more information is available for health management tasks.
- *Ice protection systems*: An IVHM system could provide an early warning of a failure in the ice protection system, thus preventing dispatch in icy conditions. Many aircraft, and especially electric aircraft, use electric resistance heating for ice protection on the air data sensors, wings, nacelles, and cockpit windshield. If an IVHM system monitored the electrical resistance, continuity, or actual temperature of these systems, the pilot could receive an early warning of an impending problem, which would necessitate a different mode of operation, including a route change.
- *Cryogenic systems*: An IVHM system could be used on future aircraft to monitor and control systems that use cryogenic fluids. These systems do not currently exist on aircraft today but may someday, based on the amount of research under way. Power electronics in cryogenic systems must work flawlessly. This means that, in addition to superbly designed systems that are extremely reliable, an IVHM system may be needed to actively monitor these systems to ensure an added level of oversight. This monitoring can be accomplished by active redundant control if the benefits versus the drawbacks (e.g., weight of multiple systems, cost, reduction in overall system reliability) can be justified, or if the monitoring is mandated by the regulators for safety purposes.

As aircraft designs evolve, the industry tends to look for opportunities to reduce cost and weight. The lower the weight of the aircraft, the lower the fuel burn. However, functionality that supports IVHM capability should not be removed during these exercises because the data will always have the potential to provide insight into an unexpected situation not previously seen.

VTOL Aircraft

For VTOL aircraft intended to transport small numbers of passengers over short distances, HUMS will continue to be mandatory to detect anomalies in critical systems, to monitor usage, and to inform the flight management computer to adapt the control strategy to ensure mission safety and success. Vibration monitoring, as well as other sensing elements, will continue to be used to discern which component is starting to degrade. Regime recognition or "usage monitoring" will assist in optimizing maintenance programs based on how the VTOL aircraft is being operated. For example, military helicopters may have lower cycle counts but undergo highly aggressive maneuvering that results in high stress loads. Private and light commercial VTOL aircraft may have higher cycle counts but operate in a more mundane environmental envelope. Industrial VTOL aircraft, such as oil and gas helicopters, may operate in a salty environment over water. Depending on the mission type, the integration of HUMS into the control strategy will vary to meet the demands for availability, optimized maintenance, and mission readiness.

- Personal air vehicles (PAVs) are likely to be all-electric or hybrid-electric VTOL designs. Entrepreneurs, start-up companies, and airframe manufacturers around the world are racing to design and deliver VTOL aircraft to support the vision of urban air mobility and mobility on-demand initiatives. In 2018, Uber Air announced plans to begin flight-testing a VTOL urban air taxi service

[1] Wikipedia, definition of "semantic reasoner."

in 2020 with plans to enter service in 2023. The Bell Nexus hybrid-electric VTOL vehicle was unveiled at the Consumer Electronics Show, or "CES," in 2019. At CES 2020, Hyundai announced a partnership with Uber Air to develop a fully electric or eVTOL aircraft that would be rechargeable in five to seven minutes. A recent study indicated that PAVs are predicted to fly an average of 20 miles per day and carry two to four passengers each flight. The result is estimated to be thousands of trips per day per city and more than 100 vehicles per city [11]. While trained pilots are likely to be flying PAVs, the dense 3D urban environment - with fixed and mobile hazards - will require more autonomy to ensure safe control and operation of the aircraft. PAVs will need to have a higher level of diagnostic and prognostic capability embedded in the active control system than transport aircraft, not only to assess the current condition and predict the future condition, but to determine if the flight can be completed successfully. To accomplish this, PAVs will likely have three levels of reconfiguration control: low-level and mid-level reconfiguration (as described previously), and high-level flight adaptation at the vehicle level. High-level flight adaption at the vehicle level means that the control laws would trade off low-priority objectives for high-priority objectives, such as landing at the next available Vertiport instead of landing at the desired Vertiport or adjusting airspeed to maintain forced separation between other PAVs.

All-electric aircraft will need to have advanced control strategies because they will be entirely "fly-by-wire" and may not have any motive force other than electricity. This means that the flight control system will likely be actuated by electromechanical actuators (EMAs), or by a hydraulic system driven by electric pumps. Historically, EMAs have not been used for primary flight control because they have a single point of failure that can result if an actuator jams. Because of this reliability concern, EMA systems must be fully redundant in order to meet the presumed certification requirements for primary flight control, which adds considerable weight to the aircraft. The flight control system will need to be aware of the health of each of the EMAs, which will be difficult due to the nonlinearity function in the fault progression of electronic components, so that the flight control system can reconfigure based on what has failed. This means that the IVHM functionality will need to be embedded into the flight control system or at least be able to inform the flight control system of the condition of the EMAs. The same argument is true for electric aircraft with propulsors. Some of the designs involve many propulsors, such as the Volocopter, which has 18. Because the aircraft will be capable of flying with one or more of the propulsors inoperative, the flight control system must be capable of maintaining stable flight with unmatched propulsors. This task could be accomplished by selectively reducing power from certain propulsors or by shutting down certain propulsors to maintain balance. The IVHM implications are clear. In fact, having an advanced health management system may be critical to the successful functioning of the aircraft, even though the impact will result in a non-agile scenario if changes are needed to the IVHM functionality.

Many all-electric and hybrid-electric aircraft depend on lithium-ion batteries. These batteries are comprised of many cells that need to be individually monitored so that the battery management system can ensure that the charging and discharging of the cells is balanced. Overcharging of a cell can lead to a short-circuit fault and thermal runaway. Having a battery health management system that can detect impending failures, and provide that information to the active control system, will be critical on these aircraft to ensure an immediate and safe landing.

Unmanned Aircraft

For aircraft intended for unmanned operations that are either remotely piloted (Level 4) or fully autonomous (Level 5) (Figure 3), the designs vary dramatically from fixed wing to VTOL, and from large to small in size. Their missions vary as well. They may be used for personal transportation, surveillance, cargo delivery, firefighting, military purposes, and so on. The first recorded use of an unmanned aircraft dates back to 1849 when Austrian forces weaponized balloons and launched them toward Venice [12]. Since then, remotely piloted aircraft have been routinely used by military forces for a variety of missions. Only recently has there been a strong interest in developing unmanned aircraft for civilian purposes. The term unmanned aircraft system (UAS) is used to describe any civil unmanned aircraft that is either fully autonomous or remotely piloted and flies within the controlled airspace.

An interesting phenomenon is occurring in the design and manufacturing of UAS vehicles: most manufacturers are working independently to develop their own designs for their aircraft. UAS designs vary depending on their intended mission, and each design has different capabilities, which increase proportionally with the cost of the aircraft. The independent development is creating challenges in the ability to transfer technology knowledge from one UAS platform to another. Complicating matters is the fact that design and certification standards for the UAS, for the most part, are nonexistent. Hence, the technology that one manufacturer is implementing will not necessarily be the same technology that another manufacturer is implementing, although the outcome and results may be the same. This phenomenon means that more emphasis on machine learning techniques will be required to identify features in the data and to recognize the health state of the components and systems, since large datasets for training algorithms will not be available. Analysts will likely need to rely on transfer learning diagnostics to bridge the gap when training data is not available for one design, but diagnostic models exist for another design that is similar [14].

FIGURE 3. **Levels of autonomy.**

Levels of Autonomy					
Level 0 No Autonomy	Level 1 Pilot-Assisted	Level 2 Partial Autonomy	Level 3 Conditional Autonomy	Level 4 High Autonomy	Level 5 Full Autonomy
Manual flight and operation	Manual flight and operation	Automated routine flights	Automated flight and operation with pilot in the loop	Automated flight and operation	Self-managed automated flight and operation
	Auto-pilot, Auto-throttle, Navigation	Auto-pilot, Auto-throttle, Navigation	Waypoint flying with adaptive safety systems	Multiple autonomous aircraft at moderate scale	Operation in dyanmic and high density air space
Visual line of sight	Safer visual line of sight	Beyond visual line of sight using waypoints	Beyond visual line of sight integration into air space	Automated support for take-off and landing	Integration of weather services
Operator knowledge and adherence to restrictions, conditions, and weather	Adherence to altitude restrictions	Automated air space authorization	Low number of autonomous aircraft	UTM integration	ATM and UTM integration
Pilot Dependency					
Trained and Skilled Pilot(s)	Trained and Skilled Pilot (s)	Trained and Skilled Pilot(s)	Single Pilot Operations plus Remote Pilot	Remote Pilot	No Pilot

Adapted from Ref. [3], Intel Corporation presentation to NASA: Building the Autonomous Future

Level 5 UAS vehicles must be capable of making decisions and taking appropriate action to ensure the integrity and safety of the aircraft for the remainder of the flight. The avionics needed for UAS vehicles will be different when compared to avionics installed on transport aircraft. Without an onboard or remote pilot to mitigate an unusual or unexpected situation, the onboard sensors will have to be highly reliable and efficiently powered. Concurrently, the systems will need to be highly redundant with control laws that are robust (capable of performing without failure under a wide range of conditions) and resilient (capable of recovering from or adjusting easily to internal or external disturbances including cyberattacks). This means that IVHM functionality (state detection, health assessment, and prognostic assessment) will be intrinsic to the control systems. These aircraft will use algorithms and artificial intelligence for situational awareness and to reconfigure control laws according to possible fault consequences. Prognostics will allow the aircraft more time to react [15]. However, because the manufacturers are not sharing data and information and the vehicle designs are very different, the process to develop autonomous algorithms will be slower than for transport aircraft where an abundance of data for training algorithms already exists. What is also unsettled is how the software needed to support Level 5 autonomy will be developed, tested, simulated, and certified within the cost constraints of the vehicle's intended purpose.

Summary

The decision to integrate IVHM capability into the active control of aircraft systems is influenced by the intended use of the aircraft. Transport aircraft are likely to have data from the IVHM system embedded in the control systems for selected components and systems, but not at the vehicle level. Rotorcraft are likely to have the data from the HUMS embedded in the control systems for critical components and systems, and at the vehicle level if the action could result in a more conservative flight condition. UAS vehicles are likely to have data from the IVHM system fully embedded in the control systems for critical components and systems as well as at the vehicle level.

Recommendations

1. When deciding what level of IVHM functionality should be designed into the active control at the component level, system level, or aircraft

level, the intended use of the aircraft should be a major consideration.

2. When designing a new aircraft, IVHM system requirements should be specified early in the design phase to ensure the capability meets the requirements for certification and redundancy in order to be used for active control.
3. To ensure data integration for onboard real-time IVHM functionality, stakeholders should align on a standard for data interoperability, data access, and data sharing.
4. To ensure the safe introduction of UAS vehicles into the national airspace, stakeholders should share lessons learned in a manner that protects their intellectual property, but also advances the development of autonomous algorithms.
5. IVHM functionality should not be removed during cost and weight reduction exercises because the data will always have the potential to provide insights into unexpected situations.

Challenges to Implementing IVHM Capabilities into Aircraft Control Systems

The integration of IVHM capability into active control loops of aircraft systems has been limited by barriers that have been put in place to ensure airworthiness, such as development assurance level (DAL) requirements and data provenance. Any technology that challenges these barriers must be proven to have no negative impact on the airworthiness of the aircraft. Further, any technology that challenges these barriers must be proven to have a positive impact, such as an economic or social benefit or a safety enhancement. If the active control is designed to use information from an IVHM system, the latter must help overcome these barriers by being able to access data from all aircraft systems, being designed to meet the certification requirements for airborne equipment, managing uncertainty in prognostic algorithms, and ensuring a harmonious integration with the pilot. The following sections discuss these challenges.

Data Interoperability for Vehicle-Level Reasoners

For an IVHM system to be effectively integrated into the active control of components, systems, or aircraft-level systems, information from multiple disparate sources must be available. Onboard data integration has been impeded by technology outpacing the lengthy aircraft development cycle and by data access and ownership issues between stakeholders. Aircraft development cycles commonly take 7-10 years while advancements in technology, in particular digital technology, follow the law of accelerating returns [16], making giant leaps every two to three years. Hence, when an aircraft enters service, the disparity between what was desired and envisioned for data integration, versus what was actually achieved, may be vastly different. During the aircraft development cycle, technical evolution may obviate the initial design before even entering service. Retrofit solutions can bridge this disparity but are expensive due to the cost of an STC. In order for an IVHM capability to be successful, access to relevant data is necessary, regardless of whether the capability resides onboard or off-board. In the last decade, the aviation industry has created self-induced barriers to data interoperability as a direct result of monetization of aircraft data by stakeholders, including "operational data ownership, intellectual property rights, data security, traceability, and governance" [17].

IVHM systems must fuse data from multiple sources as input to reasoners that are built from a variety of data-driven methods, model-based methods, and machine learning techniques. In a future world, a vehicle-level reasoning system [18] would be comprised of low-level reasoning, mid-level reasoning, and high-level reasoning. Low-level reasoning would use sensor data and diagnostic features to accurately detect the failure modes present in a component. Mid-level reasoning would use information provided by low-level reasoners to predict the health status of the systems that would be impacted by the indicted component as a fault progresses. High-level reasoning would use information provided by the low-level and mid-level reasoners to make a determination about the overall ability for the aircraft to successfully complete its intended flight or mission. Lack of data interoperability prevents the development of vehicle-level reasoners in the current world [19]. To overcome this barrier, a requirement to share data must be present from the preliminary design phase to ensure data is accessible, follow industry standards, avoid conflicts of interest concerning the use of the data (pilot unions, warranty claims, etc.), and that the underlying system has adequate sensing to ensure accuracy and low ambiguity of the reasoning functions.

Having vehicle-level reasoners will be essential for PAVs and Level 5 UAS vehicles in general as well as for special programs such as the military's future vertical lift fleet modernization initiative. For multi-aircraft military operations, if a vehicle-level reasoner determines that the mission cannot be completed as planned, the reasoner can suggest how the aircraft can be re-tasked, such as swapping one aircraft for another.

Several SAE International committees maintain standards pertaining to data interoperability and data exchange for IVHM and HUMS. In 1997, the Rotorcraft Industry Technology Association launched an initiative to develop consensus standards around HUMS, working with representatives from Boeing, Bell Helicopter, Goodrich, Sikorsky

Aircraft, and Smiths. Some data interoperability standards associated with IVHM and HUMS are

- SAE AIR6904™ defines guidelines for creating a framework of data interoperability for health management in aerospace.
- SAE JA6268™ defines a methodology for exchanging data from health-ready components during design and run-time and defines five levels of IVHM capability.
- SAE AS5395™ defines a standard for data interchange for onboard and off-board HUMS.

SAE-ITC-an organization comprised of public and private entities collaborating in a neutral forum to drive innovative solutions to key industry challenges-launched the Mobility Data Collaborative in December 2019. The goal of this group is to create a framework to improve data sharing between urban areas and PAVs as these vehicles are integrated into the airspace. The collaborative is expected to drive the development of standards for data interoperability for PAVs and UAS over the next few years.

Certification

The certification requirements for airborne hardware and software are stringent, which is necessary to ensure the safe operation of the aircraft. SAE ARP4754™ and ARP4761™ provide guidelines in the context of FAA 14CFR25 and EASA CS-25 for developing systems supporting aircraft-level functions that have failure modes with the potential to affect safety of the aircraft. Any introduction of IVHM capability into the active control of airborne systems would be required to meet the guidelines outlined in these and other regulatory approved certification documents.

During the requirements definition phase of a new aircraft design, the requirements - for safety - are kept separate from the requirements for performance. For example, it is okay for a vehicle to run out of fuel as long as the vehicle does not stop in an unsafe situation. Since IVHM has been considered a means to optimize performance and reduce cost rather than a means to enhance safety, the information provided from the IVHM system has not been used in the System Safety Assessment for any known aircraft [20]. Because IVHM has been perceived to have no effect on the safety of the aircraft, most IVHM systems are classified as DAL D or E (see Tables 1 and 2). Control systems that have an effect on safety are assigned a higher DAL, depending on the consequence of failure. The methodologies used to assess the DAL include failure modes and effects analysis, fault tree analysis, hazard and operability studies, preliminary hazards analysis, and common cause analysis. While each of these methodologies are effective, they are also based on historical statistical information, and the analysis is rarely updated as the aircraft matures. These methodologies could be enhanced by using IVHM data from similar aircraft designs to supplement the initial analysis. Further, the hardware and software elements that comprise the IVHM system would need to be designed to meet redundancy and low-ambiguity requirements to ensure the ongoing use of the data for analysis and control.

TABLE 1. Top-level Functional Development Assurance Level (FDAL) assignment.

Top-level failure condition severity classification	Associated top-level FDAL assignment
Catastrophic	A
Hazardous/severe major	B
Major	C
Minor	D
No safety effect	E

Examples exist today of EHM systems and HUMS that have been certified to DAL B. These systems have achieved this assurance level through early incorporation into the aircraft design and by redundancy of hardware (e.g., sensors) and software (e.g., channels, partitions, robust models). Initially, some engine manufacturers embedded EHM functionality into the engine control software in order to save cost and weight. However, the engine manufacturers found that the EHM functionality could not be certified to the same DAL as the safety-critical engine control software (due to the uncertainty in the EHM health assessments) without undergoing tremendous cost. The solution was to partition the engine control software such that the EHM functionality and the control functionality were completely separate. In time, the engine manufacturers realized that more flexibility was needed for developing and deploying diagnostic and prognostic algorithms, which drove some platforms to move the EHM functionality to a separate device mounted on the engine.

Further, since the mid-1980s, engine manufacturers have been certifying engine control software that uses engine models, similar to a digital twin, that utilize indirect engine measurements, or virtual sensors. These virtual sensors are used by the engine controllers as a redundant source of information as well as a substitute

TABLE 2. Software assurance levels for transport aircraft.

Failure condition	Onboard (DO-178C)	Ground-based (DO-278A)
Catastrophic	Level A	AL1
Hazardous/severe	Level B	AL2
Major	Level C	AL3
Minor	Level D	AL5
No effect	Level E	AL6

FIGURE 4. NASA MBCPHM approach.

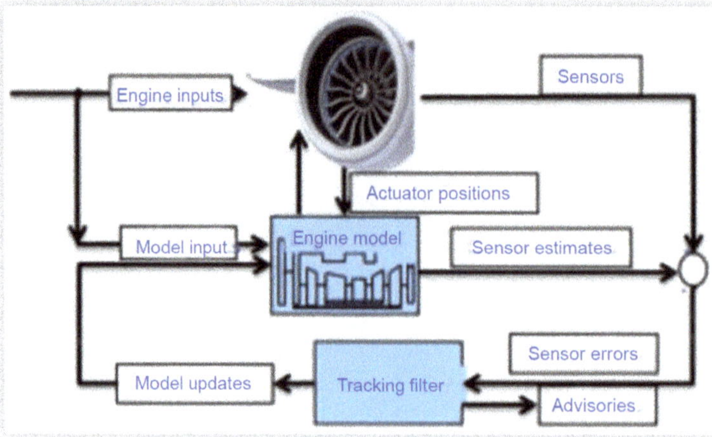

sensor when physical sensors are faulty or do not exist. Because of the uncertainty associated with these virtual sensors, the engine manufacturers include conservatism into the engine control laws to ensure minimum performance is met during all scenarios. Recently, research aimed at reducing the conservatism in engine controls has been conducted by NASA and the engine and airframe manufacturers. This research uses a model-based control PHM (MBCPHM) approach to develop new control laws that are tailored to the prognostic health assessment and usage-based lifing of a specific engine (Figure 4) [22].

For VTOL aircraft, the certification requirements are the same as for transport aircraft. This means that most PAVs will also need to meet these same certification requirements. Since these aircraft are expected to have more autonomy than transport aircraft, implementation of IVHM functionality should be considered a safety requirement and treated as such throughout the hardware and software verification and validation process.

For UAS vehicles, very few certification standards currently exist. Since these aircraft are expected to be Level 4 or Level 5 autonomous vehicles, implementation of IVHM functionality must be considered a safety requirement and treated as such throughout the hardware and software verification and validation process. Recently, the SAE S-18 Aircraft and System Development and Safety Assessment Committee joined with the SAE AS-4 Unmanned Systems Steering Committee to form a new task group called S-18UAS. This new task group consists of industry experts who are collaborating to write several consensus standards and recommended practices for UAS operations. The task group is examining the applicability of ARP4754 and ARP4761 to unmanned aircraft to identify shortcomings and gaps. Once these documents are published, they will likely become the approved guidance for certifying autonomous aircraft under the Code of Federal Regulations (CFRs). As of the writing of this SAE EDGE™ Research Report, the task group had not published any documents.

Uncertainty in Prognostics Many in the industry believe that IVHM is about management, not about control. The information provided by an IVHM system can be used to inform the control system so that the fault consequences can be better managed-potentially extending a failed operational situation from becoming a fail-degraded situation. Some industry analysts believe that trending data is prognostics. Trending can be used to detect changes in a particular component, or it can be extended to a fleet to identify outliers that potentially indicate an emerging trend. Merely trending data to estimate RUL based on historical data is not really prognostics. There are several interpretations of what is included in prognostics, such as estimating the mean time to failure (based on statistical analysis of a large population of the target component), estimating the health state of the target component using sensor data, and estimating the RUL of the target component using algorithms that are customized for the failure mode detected. The bases for these estimations are quite diverse, which also means that predictions must be interpreted differently because the uncertainty is quite distinct.

Injecting information into the control system requires that the input be highly robust and trustworthy. Redundant hardware and software will partially meet this demand. The larger concern is how to quantify and manage the uncertainty in the prognostic assessments, which is present in any method used. IVHM practices [23] depend on accurate and robust detection of incipient failures and accurate estimation of the RUL of the failing component or system. Predicting failure progression and the time to functional failure has inherent uncertainty in part because the aircraft operates in an evolving ecosystem where the future usage can only be assumed. Additionally, no two aircraft are identical, and even similarly equipped aircraft are operated in varied environments and under different operational parameters. This variance negatively impacts the consistency of data needed for statistical analysis and health state evaluation.

FIGURE 5. **Uncertainty management methodology.**

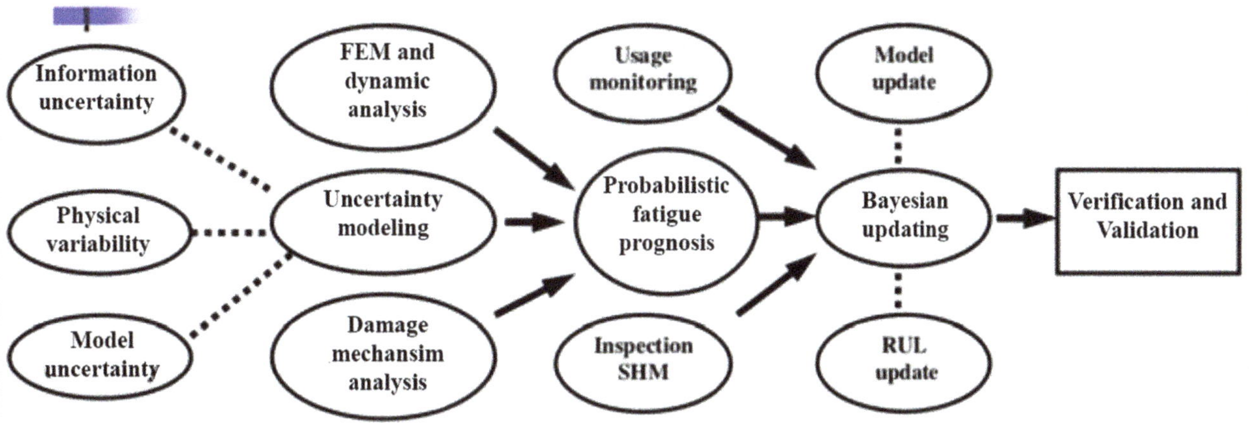

To manage uncertainty, begin by first focusing on accurate sensing, modeling, and data fusion for "simple" components or systems, and then expanding the techniques to more complex systems, such as the flight control system. The utilization of big data analytic techniques has resulted in the accumulation of ground-truth data for training algorithms, which has enabled machine learning techniques to be used to improve modeling accuracy. Rigorous data fusion techniques have allowed information from other sources to be merged with sensed data to reduce uncertainties [24]. Leveraging other data sources (i.e., weather) and different approaches, such as deterministic and probabilistic models, has been shown to reduce uncertainty in prognostics. Figure 5 illustrates an example methodology for managing the uncertainty in prognostics.

Human-Machine Teaming SAE AIR6915 examines the role that IVHM could play in improving the human-machine interface between the pilot and the aircraft. The human-machine interface involves the information that is provided from the aircraft to the pilot, the response from the pilot to that information, and the feedback from the aircraft to the pilot's response. Human Factors Systems Engineering strives to ensure that, not only is the information provided to the pilot accurate and concise, but actionable. Unfortunately, many accidents have resulted from pilots taking the wrong course of action after being overwhelmed with information.

Ideally, pilots and engineers would collaborate during the design phase to ensure the priority of information displayed to the pilot is accurate and not overwhelming. Historically, flight test pilots have been some of the best and most experienced pilots in the industry. These pilots are not representative of the worldwide pool of pilots who will be flying the new aircraft. Hence, the testing of any new aircraft design would benefit by including less-experienced pilots. Perhaps this could occur using a simulator rather than the cockpit, to test pilot response to the aircraft information displayed during an emergency situation. It could also test their ability to override the nominal aircraft response to an unusual or unexpected situation.

More authority should be given to the flight control systems of autonomous aircraft because they are likely to be operated without a pilot onboard or by pilots with less skills than trained pilots flying transport or military aircraft. As aircraft become more autonomous, the pilot's ability to independently respond to fault scenarios degrades. Hence, the pilot must remain engaged in all phases of flight in order to be able to react quickly to mitigate a situation and take control of the aircraft. Further, the autonomous system must also ensure a harmonious interface with the pilot through appropriate notifications so that the pilot is not surprised by any automated actions taken by the system. An example of failure to achieve this harmonious interface is the Boeing 737 MAX where the pilots were unable to recover control after the Maneuvering Characteristics Augmentation System automatically and repeatedly forced the aircraft into a nosedive, resulting in the crash of Lion Air Flight 610 and Ethiopian Airlines Flight 302. An earlier example of an interface failure occurred in 1994 when China Airlines Flight 140 crashed at Nagoya Airport after the crew was unable to overcome the autopilot once the Take-off/Go-Around mode was inadvertently selected.

IVHM information embedded in the active control systems could provide the pilot with enough advanced warning to allow more time to react to an unusual or unexpected situation. The hand-over from the aircraft to the pilot must be well coordinated so incidents are avoided. Vehicle-level reasoning could help mitigate human-machine response errors and augment the overall safety of the aircraft. This could be achieved by evaluating the sensor inputs, detecting and prioritizing faults, predicting the consequences of the fault, alerting the pilot with the most important information, and informing the relevant control system to take automated action. The overall workload of the pilot could be reduced through the use of these reasoners.

Ejection seat systems have saved the lives of hundreds of military pilots. History has shown that the likelihood of a pilot safely ejecting is highly dependent on how quickly he or she responds to the events unfolding. The aircraft control systems and vehicle-level reasoners could assess the unusual and unexpected situation more quickly and accurately than the pilot could, thus providing the pilot with a warning to eject earlier.

Pilots must be trained to understand the aircraft's automated response to disturbances and how best to override the response, as needed. The May 2018 United Kingdom Ministry of Defence report concluded that "Human mental capacity is increasingly unable to cope with the data deluge involved in conflict, while computer algorithms are challenged by uncertainty and ambiguity in data and decision-making. Optimising human and machine capabilities in teams that maximise strengths and mitigate weaknesses is essential" [25].

The following examples are intended to trigger some thought surrounding whether IVHM functionality could have made a difference in pilot response and the outcome of the event. These examples are not intended to imply a shortcoming of the aircraft or pilot training.

- In 1999, a Learjet 35 carrying pro-golfer Payne Stewart, three associates, and two pilots crashed hours after the aircraft lost cabin pressurization, incapacitating everyone onboard. Once the aircraft deviated from its planned flight route, air traffic control attempted to contact the crew multiple times without success. Multiple interceptions indicated that the cabin was dark and the cockpit windshield was opaque [26]. Could IVHM functionality have detected and predicted the failure that lead to decompression event? Could IVHM functionality have informed the flight control system to initiate an immediate descent to a lower altitude where the flight crew could have been revived?
- In 2009, Air France Flight 447 crashed into the Atlantic Ocean after the A330's air data pitot tubes iced over in-flight, causing the airspeed indication to report erroneously and leading to autopilot disconnect. The pilots failed to recognize the instrumentation error and inadvertently took conflicting actions that resulted in the aircraft stalling, rapidly descending from 38,000 ft, and crashing into the ocean at full engine throttle. The findings pointed to multiple instances of inappropriate pilot response to the situation where they did not follow standard operating practices, which ultimately led to the crash. Could IVHM functionality have detected the airspeed discrepancy, accommodated for the failure, and provided only the critical information needed to the pilots to safely fly the aircraft?
- In 2010, Qantas Flight 32 experienced an uncontained Engine 2 failure shortly after take-off. The shrapnel from the failed engine severed hydraulic lines and fuel lines, disabling the hydraulic system, causing a fuel tank fire, damaging the nacelles and landing flaps, and impacting the engine control, the landing gear, and the flight control system. The pilots were able to quickly determine that the aircraft was controllable; however, the assessment of the aircraft's status took over two hours as the pilots followed standard operating practices and checklists. The aircraft remained in a holding pattern until the assessment was complete and a plan to land the aircraft safely could be developed. Could IVHM functionality have evaluated the faults and sensor readings more quickly than the time it took for the pilots to perform the assessment, reducing the potential risk of further complications?
- In 2015, TransAsia Flight 235 crashed shortly after take-off when a fault in the Engine 2 auto-feather unit resulted in the aircraft's automated take-off power control system to auto-feather the engine. In response, the pilot-in-command mistakenly pulled back the throttle on Engine 1, which resulted in the aircraft losing altitude and crashing into a river. Could IVHM functionality have provided additional capability or information that would have prevented the pilot from shutting down the wrong engine?
- While aircraft accidents caused by malicious intent are rare, could an onboard IVHM functionality that uses biometric sensors detect when a pilot is in distress and take action to alert ground operations before the flight departs?

For every accident caused by pilot error, there are many more instances where the pilot was able to overcome the disturbance and safely fly the aircraft. Artificial intelligence is not likely to replace the pilot on transport aircraft in the foreseeable future due to the perception that passengers feel safer with a pilot onboard to mitigate any unusual or unexpected situation. Machine learning is a long way from being able to replace the pilot's ability to quickly assess the situation when an untoward event occurs. At what point do we stop trusting the pilot and start trusting the aircraft systems first? As aircraft become more complex and more autonomous, the need for human-machine teaming will become even more critical. Machines will continue to do what they do best: crunch data quickly. Humans will continue to do what they do best: make decisions based on the information provided. The teaming of the human and the machine can be improved through the use of IVHM data embedded in the active control system.

Summary

The barriers that have limited the integration of data from IVHM systems into the active control of aircraft systems can be addressed by resolving data access and ownership issues, designing IVHM systems to meet the certification requirements for airborne equipment, managing the uncertainty in prognostic algorithms, and leveraging the best of humans and machines to enhance the decision-making process.

Recommendations

1. The industry should reach a consensus on standards for data interoperability that allows for data access and data sharing while protecting intellectual property.
2. Systems Engineers should include IVHM capability in the requirements definition phase of new designs to ensure that hardware and software redundancy is achieved for DAL A or DAL B certification.
3. Uncertainty in prognostics should be managed by focusing on simple systems first and then expanding the proven techniques to more complex systems.
4. Uncertainty in prognostics should also be managed through the accumulation of ground-truth data for training, rigorous data fusion and testing, and leveraging varied approaches.
5. IVHM reasoners should be used to augment human machine teaming to ensure the pilot is provided with only the most important and necessary information to make a correct decision and take action.
6. More authority should be given to the flight control systems of autonomous aircraft because they are likely to be operated in dense 3D urban environments with fixed and mobile hazards.

Summary

As aircraft have become more technologically capable in terms of sensing and processing data, the use of IVHM data for maintenance optimization has expanded to other systems beyond engines and APUs. The need for large transport aircraft will continue to exist alongside the desire for more personal mobility, such as VTOL aircraft and PAVs. Autonomous systems will fill the void created by a shortage of skilled pilots and will offer economical solutions for transporting small numbers of passengers, delivering cargo, and performing unique tasks, such as fighting fires. The development of next-generation military aircraft will accelerate the advancement of technologies that enable autonomy as well as IVHM functionality. The use of IVHM for active control in addition to maintenance optimization has been discussed but not implemented due to barriers regarding data provenance, designing IVHM systems to meet certification requirements for airborne equipment, managing uncertainty in prognostic algorithms, and leveraging the best of humans and machines to enhance the decision-making process. The intended use of an aircraft will greatly influence the level to which these barriers would need to be addressed. Transport aircraft are likely to have data from the IVHM system embedded in the control systems for selected components and systems, but not at the vehicle level. Rotorcraft are likely to have the data from HUMS embedded in the control systems for critical components and systems and at the vehicle level, if the action could result in a more conservative flight condition. UAS vehicles are likely to have data from the IVHM system fully embedded in the control systems for critical components and systems, as well as at the vehicle level.

SAE EDGE™ Research Reports

SAE EDGE™ Research Reports, like the present report on “Unsettled Topics on the Use of IVHM in the Active Control Loop,” are intended to push further into still unsettled areas of technology of interest to the mobility industry. SAE launches these reports before attempting to form a joint working group, a cooperative research program, or a standards committee.

SAE EDGE™ Research Reports are intended to be quick, concise overviews of major unsettled areas where vital new technologies are emerging. An unsettled area is characterized more by confusion and controversy than established order. Early practitioners must confront an absence of agreement. Their challenge is often not to seize the high ground but to find common ground. These scouting reports from the frontiers of investigation are intended merely to begin the process of sorting through critical issues, contributing to a better understanding of key problems, and providing helpful suggestions about possible next steps and avenues of investigation.

SAE EDGE™ Research Reports, therefore, are fundamentally distinct from the more formal working groups approach and far removed from the more mature research program and standard’s development process.

Next Steps for the Use of IVHM in the Active Control Loop

This publication should be considered only as a first step toward clarifying the issues around the use of IVHM in the active control loop. The intention behind this and other SAE EDGE™ Research Reports is to start a dialogue among interested parties on important industry-wide topics that require further attention. The expectation is that these explorations of unsettled areas of technology will lead to the formation of working groups and, ultimately, committees that can address and resolve the issues they raise, producing a framework for developing a common vocabulary of definitions, best practices, protocols, and standards needed to support continued progress toward safer and more innovative automotive products.

The experts' collaboration that gave rise to this publication demonstrated a great willingness on the part of the industry to define the terminology, procedures, and eventually the standards needed to enable IVHM technology to move ahead as quickly and efficiently as possible. SAE International has demonstrated its lead in this and closely related areas by the SAE HM-1 Committee for Integrated Vehicle Health Management.

This SAE EDGE™ Research Report on IVHM in the active control loop identifies the following key topics for further pursuit, both through continued informal discussions among industry practitioners and through more formal working groups:

- Robust and Resilient Technologies
- Standardization
- Certification
- Human-Machine Teaming

Recommendations

The overall recommendations of this SAE EDGE™ Research Report can be summarized as follows:

1. When deciding what level of IVHM functionality should be designed into the active control at the component level, system level, or aircraft level, the intended use of the aircraft should be a major consideration.
2. When designing a new aircraft, IVHM system requirements should be specified early in the design phase to ensure the capability meets the requirements for DAL A or DAL B certification and redundancy in order to be used for active control.
3. IVHM functionality should not be removed during cost and weight reduction exercises because the data will always have the potential to provide insights into unexpected situations.
4. To ensure data integration for onboard real-time IVHM functionality, the industry should align on consensus standards for data interoperability that allow for data access and data sharing while protecting intellectual property.
5. To ensure the safe introduction of UAS vehicles into the national airspace, stakeholders should share lessons learned in a manner that protects their intellectual property but also advance the development of autonomous algorithms.
6. Uncertainty in prognostics should be managed by focusing on simple systems first and then expanding the proven techniques to more complex systems.
7. Uncertainty in prognostics should also be managed through the accumulation of ground-truth data for training, rigorous data fusion and testing, and leveraging varied approaches.
8. IVHM reasoners should be used to augment human machine teaming to ensure the pilot is provided with only the most important and necessary information to make a correct decision and take action.
9. More authority should be given to the flight control systems of autonomous aircraft because they are likely to be operated in dense 3D urban environments with fixed and mobile hazards.

Acronyms

APU - Auxiliary Power Unit

ARP - Aerospace Recommended Practice

BIT - Built-In Test

CBM - Condition-Based Maintenance

CFR - Code of Federal Regulations

DAL - Development Assurance Level

EASA - European Union Aviation Safety Agency

EHM - Engine Health Management

EMA - Electromechanical Actuator

ETOPS - Extended Operations

eVTOL - Electric Vertical Take-off and Landing

FAA - Federal Aviation Administration

FDAL - Functional Development Assurance Level

FDIA - Fault Detection, Isolation, and Accommodation

HUMS - Health and Usage Monitoring System

ISO - International Standards Organization

IVHM - Integrated Vehicle Health Management

MBCPHM - Model-Based Control and Prognostics and Health Management

MCAS - Maneuvering Characteristics Augmentation System

MRO - Maintenance, Repair, and Overhaul

NASA - National Aeronautics and Space Administration

PAV - Personal Air Vehicle

PHM - Prognostics and Health Management

RUL - Remaining Useful Life

SATAA - Sense, Acquire, Transfer, Analyze, Act

STC - Supplemental Type Certificate

UAS - Unmanned Aircraft Systems

UAV - Unmanned Aerial Vehicle

VTOL - Vertical Take-Off and Landing

Acknowledgements

Recognition should go first to all the participants, all of whom also provided feedback on the draft version of this publication. Without their input and initiative, this SAE EDGE™ Research Report would not have been possible.

Dr. Shreeder Adibhatla, *General Electric Aviation*
Carina Birgersdotter, *SAAB, retired*
Andy Hess, *The Hess PHM Group*
Dr. Ian Jennions, *Cranfield University*
Dave Larsen, *Collins Aerospace*
Linda O'Flarity, *Pratt & Whitney*
Steve O'Flarity, *Pratt & Whitney*
Dr. Ravi Rajamani, *drR2 consulting*
Douglas Felipe Rodrigues da Silva, *Embraer S.A.*
Dr. Antonios Tsourdos, *Cranfield University*
Dr. George Vachtsevanos, *Georgia Institute of Technology, retired*

The authors of this document together with the SAE Team responsible for its creation join in expressing our deepest appreciation to all the individuals mentioned above.

Rhonda D. Walthall
Collins Aerospace

Bibliographic References

1. https://www.faa.gov/data_research/aviation/aerospace_forecasts/media/FY2019-39_FAA_Aerospace_Forecast.pdf.
2. SAE International, "IVHM Concepts, Technology, and Implementation Overview," SAE Standard ARP6803™, 2016.
3. SAE International, *Integrated Vehicle Health Management - System of Systems Integration* (Warrendale, PA: SAE, 2017). ISBN:978-0-7680-8431-3.
4. SAE International, "AIR5871A™ - Prognostics for Aerospace Propulsion Systems."
5. SAE International, "JA6268™ - Design & Run-Time Information Exchange for Health-Ready Components."
6. Day, J. and Harding, P., "The Status of HUMS for Helicopters within the UK Ministry of Defence," 2017.
7. FAA Miscellaneous Guidance, "MG-15: Airworthiness Approval of Rotorcraft Health Usage Monitoring System (HUMS)," AC 29-2C Certification of Normal Category Rotorcraft, 2018.
8. https://arc.aiaa.org/doi/pdf/10.2514/6.1988-4039.
9. Witwer, B., "Systems Integration of the 777 Airplane Information Management System (AIMS)," *IEEE Aerospace and Electronic Systems Magazine* 11(4):17-21, 1996.
10. https://www.theatlantic.com/technology/archive/2016/03/has-the-self-flying-plane-arrived/472005/.
11. Guiliano, A. and NASA, "Our Vision on Urban Air Mobility: Challenges and Role of UTM," 2019.
12. https://en.wikipedia.org/wiki/Unmanned_aerial_vehicle.
13. Nanduri, A., "Building the Autonomous Future," in *Intel, NASA Workshop*, Aug. 2019, NASA Ames Research Center, Moffett Field, CA.
14. Lei, Y., Yang, B., Jiang, X., Jia, F. et al., "Applications of Machine Learning to Machine Fault Diagnosis: A Review and Roadmap," *Mechanical Systems and Signal Processing* 138:106587, 2019.
15. Pelham, J., Fan, I., Jennions, I., and McFeat, J., "Dual Use IVHM for UAS Health Management," SAE Technical Paper 2013-01-2202, 2013, https://doi.org/10.4271/2013-01-2202.
16. https://www.kurzweilai.net/the-law-of-accelerating-returns.
17. Rajamani, R., Whitfield, M., and Ravi Kumar, G.V.V., "Data Interoperability for Aerospace IVHM Systems," SAE Technical Paper 2019-01-1342, 2019, https://doi.org/10.4271/2019-01-1342.
18. Roemer, M., "A Hierarchical Reasoning Structure to Support Aerospace IVHM," *SAE Int. J. Aerosp.* 4(2):1176-1183, 2011, https://doi.org/10.4271/2011-01-2665.
19. Khan, F., Jennions, I., and Sreenuch, T., "Integration Issues for Vehicle Level Distributed Diagnostic Reasoners," SAE Technical Paper 2013-01-2294, 2013, https://doi.org/10.4271/2013-01-2294.
20. Lin, Y., Skaf, Z., and Jennions, I., "A Survey on Operational Safety Assessment in the Aviation Industry and Its Link to IVHM," SAE Technical Paper 2012-01-2590, 2015, https://doi.org/10.4271/2015-01-2590.
21. SAE International, "Guidelines for Development of Civil Aircraft and Systems," SAE Standard ARP4754A, 2010.

22. Adibhatla, S. et al., "Propulsion Control Technology Development Needs to Address NASA Aeronautics Research Mission Goals for Thrusts 3a and 4," in *AIAA Propulsion and Energy Forum*, Cincinnati, OH, 2018.
23. Vachtsevanos, G. and Valavanis, K., "A Novel Approach to Integrated Vehicle Health Management (IVHM)," in *Science and Technology Organization Applied Vehicle Panel, AVT-305 Specialist Meeting on "Sensing Systems for Integrated Vehicle Health Management for Military Vehicles,"* Brno, Czech Republic. May 2019.
24. Liu, Y., "Uncertainty Management in Prognostics," 2010.
25. https://assets.publishing.service.gov.uk/government/uploads/system/uploads/attachment_data/file/709359/20180517-concepts_uk_human_machine_teaming_jcn_1_18.pdf.
26. https://en.wikipedia.org/wiki/1999_South_Dakota_Learjet_crash.

Contact Information

EDGEresearch@sae.org.

The authors of this document together with the SAE Team responsible for its creation join in expressing our deepest appreciation to all of the individuals who contributed.

www.ingramcontent.com/pod-product-compliance
Lightning Source LLC
LaVergne TN
LVHW070912210926
848482LV00017B/1063
* 9 7 8 1 4 6 8 6 0 1 8 5 5 *